2021年

全国水利发展统计公报

2021 Statistic Bulletin
on China Water Activities

中华人民共和国水利部 编

Ministry of Water Resources, People's Republic of China

·北京·

图书在版编目(CIP)数据

2021年全国水利发展统计公报 = 2021 Statistic Bulletin on China Water Activities / 中华人民共和国水利部编. -- 北京：中国水利水电出版社，2022.12
ISBN 978-7-5226-1368-0

Ⅰ.①2… Ⅱ.①中… Ⅲ.①水利建设-经济发展-中国-2021 Ⅳ.①F426.9

中国版本图书馆CIP数据核字(2022)第256514号

书　　名	2021年全国水利发展统计公报 2021 Statistic Bulletin on China Water Activities 2021 NIAN QUANGUO SHUILI FAZHAN TONGJI GONGBAO
作　　者	中华人民共和国水利部　编 Ministry of Water Resources, People's Republic of China
出版发行	中国水利水电出版社 (北京市海淀区玉渊潭南路1号D座　100038) 网址：www.waterpub.com.cn E-mail：sales@mwr.gov.cn 电话：(010) 68545888 (营销中心)
经　　售	北京科水图书销售有限公司 电话：(010) 68545874、63202643 全国各地新华书店和相关出版物销售网点
排　　版	中国水利水电出版社微机排版中心
印　　刷	河北鑫彩博图印刷有限公司
规　　格	210mm×297mm　16开本　4印张　65千字
版　　次	2022年12月第1版　2022年12月第1次印刷
印　　数	0001—1000册
定　　价	**39.00元**

凡购买我社图书，如有缺页、倒页、脱页的，本社营销中心负责调换

版权所有·侵权必究

目 录

1 水利固定资产投资 …………………………………… 1
2 重点水利建设 ………………………………………… 4
3 主要水利工程设施 …………………………………… 8
4 水资源节约利用与保护 ……………………………… 13
5 防汛抗旱 ……………………………………………… 15
6 水利改革与管理 ……………………………………… 17
7 水利行业状况 ………………………………………… 26

Contents

I. Investment in Fixed Assets 30

II. Key Water Projects Construction 34

III. Key Water Structures and Facilities 37

IV. Water Resources Conservation, Utilization and Protection 42

V. Flood Control and Drought Relief 43

VI. Water Management and Reform 45

VII. Current Status of the Water Sector 54

　　2021年是中国共产党成立100周年,是党和国家历史上具有里程碑意义的一年。一年来,在党中央、国务院的坚强领导下,各级水利部门心怀"国之大者",完整、准确、全面贯彻新发展理念,深入落实习近平总书记"节水优先、空间均衡、系统治理、两手发力"治水思路和关于治水重要讲话指示批示精神,真抓实干、克难奋进,推动新阶段水利高质量发展迈出有力步伐,实现了"十四五"良好开局。

1 水利固定资产投资

2021年，水利建设完成投资7576.0亿元，其中：建筑工程完成投资5851.3亿元，占77.2%；安装工程完成投资330.1亿元，占4.4%；机电设备及工器具购置完成投资203.6亿元，占2.7%；其他完成投资（包括移民征地补偿等）1191.0亿元，占15.7%。

	2014年/亿元	2015年/亿元	2016年/亿元	2017年/亿元	2018年/亿元	2019年/亿元	2020年/亿元	2021年/亿元
全年完成	4083.1	5452.2	6099.6	7132.4	6602.6	6711.7	8181.7	7576.0
建筑工程	3086.4	4150.8	4422.0	5069.7	4877.2	4987.9	6014.9	5851.3
安装工程	185.0	228.8	254.5	265.8	280.9	243.1	319.7	330.1
机电设备及工器具购置	206.1	198.7	172.8	211.7	214.4	221.1	250.0	203.6
其他（包括移民征地补偿等）	605.6	873.9	1250.3	1585.2	1230.1	1259.7	1597.1	1191.0

在全年完成投资中，防洪工程建设完成投资2497.0亿元，占33.0%；水资源工程建设完成投资2866.4亿元，占37.8%；水土保持及生态工程建设完成投资1123.6亿元，占14.8%；水电、机构能力建设等专项工程完成投资1088.9亿元，占14.4%。

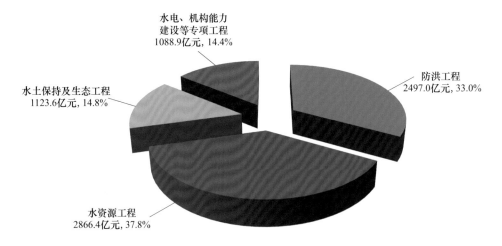

2021年分用途完成投资

七大流域完成投资6112.4亿元，东南诸河、西北诸河以及西南诸河等其他流域完成投资1463.6亿元；东部、中部、西部、东北地区完成投资分别为3166.0亿元、1855.6亿元、2320.4亿元和234.0亿元。

在全年完成投资中，中央项目完成投资67.8亿元，地方项目完成投资7508.2亿元。大中型项目完成投资1690.1亿元，小型及其他项目完成投资5885.9亿元。各类新建工程完成投资5746.0亿元，扩建、改建等项目完成投资1830.0亿元。

全年水利建设新增固定资产4017.8亿元。截至2021年年底，在建项目累计完成投资18350.1亿元，投资完成率为62.2%；累计新增固定资产9888.4亿元，固定资产形成率为53.0%，比上年增加3.6个百分点。

当年在建的水利建设项目31614个，在建项目投资总规模29502.1亿元，较上年减少7.0%。其中：有中央投资的水利建设项目15936个，较上年减少2.3%；在建投资规模12553.2亿元，较上年减少5.7%。新开工项目20900个，较上年减少7.2%，新增投资规模6664.3亿元，比上年减少16.0%。全年水利建设完成土方、石方和混凝土方分别为26.8亿立方米、3.8亿立方米、2.4亿立方米。截至2021年年底，在建项目计划实物工程量完成率分别为：土方97.5%、石方98.9%、混凝土方89.6%。

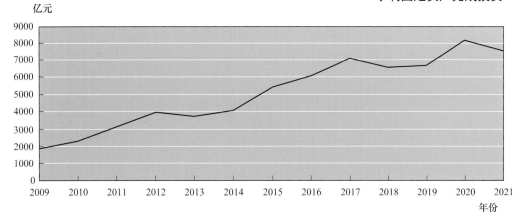

水利固定资产完成投资

2 重点水利建设

江河湖泊治理。2021年，在建江河治理工程4903处，其中：堤防建设643处，大江大河及重要支流治理713处，中小河流治理2928处，行蓄洪区安全建设及其他项目619处。截至2021年年底，在建项目累计完成投资3916.7亿元，项目投资完成率67.0%。长江中下游河势控制和河道整治9项工程已开工9项，全部建成并发挥效益；黄河下游防洪2项工程已开工2项，全部建成并发挥效益；进一步治淮38项工程已开工34项，其中16项建成并发挥效益；洞庭湖、鄱阳湖治理工程5项工程已开工5项，其中3项建成并发挥效益；太湖流域水环境综合治理12项工程已开工12项，其中8项已建成并发挥效益。

水库及枢纽工程建设。2021年，在建水库及枢纽工程1275座。截至2021年年底，在建项目累计完成投资3536.8亿元，项目投资完成率67.5%。河南袁湾水库、贵州观音水库等大型水库及枢纽和溪尾水库等19座中型水库开工。内蒙古东台子水库、陕西东庄水利枢纽、海南迈湾水利枢纽等工程实现年度导截流目标；安徽江巷水库、贵州黄家湾水库、云南德厚水库、江西四方井水利枢纽等工程等下闸蓄水。

三峡后续工作规划实施管理。2021年度，中央安排国家重大水利工程建设基金（三峡后续工作）预算113.06亿元，较2020年增加35.16%，其中地方转移支付109.94亿元、中央部门（单位）3.12亿元。当年完成投资92.29亿元，占预算的81.63%。全年在三峡库区及长江中下游重点影响区共实施移民安稳致富和促进库区经济社会发展项目377个，产业发展项目直接受益和促进就业移民群众7.5万人，对三峡库区移民子女2857人就读高职进行补助，劳动力技能培训和就业扶持6215人，城镇帮扶项目移民受益人数37.96万人、农村帮扶项目移民受益人数32.19万人；实施库区生态环境建设与保护项目120个，库岸环境综合整治长度140.18公里、支流系统治理31条；实施库区地质灾害防治项目120个，地质灾害隐患治理66处、避险人数2190人，高切坡监测预警3091处、高切坡监测保护受影响居民人数46.04万人；实施三峡工程运行对长江中下游重点影响区影响处理项目31个，长江中下游影响区护岸长度90.08公里；实施三峡工程综合管理能力建设和综合效益拓展项目31个，为推动新阶段三峡后续工作高质量发展提供了支撑。

水资源配置工程建设。 2021 年，水资源配置工程在建投资规模 6911.9 亿元，累计完成投资 4331.4 亿元，项目投资完成率 62.7%。陕西引汉济渭二期，海南琼西北供水等工程开工建设；引江济淮、滇中引水、珠江三角洲水资源配置等工程加快实施。

农村水利建设。 2021 年，全国共完成农村供水工程建设资金 525 亿元，提升 4263 万农村人口供水保障水平。当年下达中央预算内投资 75.4 亿元用于大型灌区续建配套与现代化改造，安排中央财政水利发展资金 70 亿元用于中型灌区续建配套与节水改造。全年新增耕地灌溉面积 1114 千公顷。截至 2021 年年底，全国自来水普及率达到 84%。

农村水电建设。 2021 年，全国农村水电建设完成投资 33.2 亿元，新增水电站 61 座，新增装机 31.2 万千瓦。

水土流失治理。 2021 年，水土保持及生态工程在建投资规模 3205.7 亿元，累计完成投资 1741.6 亿元。全国新增水土流失综合治理面积 6.8 万平方公里，其中国家水土保持重点工程新增水土流失治理面积 1.28 万平方公里。对 556 座黄土高原淤地坝进行了除险加固，整治坡耕地面积 86 万亩，新建淤地坝（拦沙坝）684 座。

行业能力建设。 2021 年，水利行业能力建设完成投资 60.1 亿元，其中：防汛通信设施投资 5.1 亿元，水文建设投资 23.8 亿元，科研教育设施投资 1.8 亿元，其他投资 29.4 亿元。

3 主要水利工程设施

水库和枢纽。全国已建成各类水库97036座,水库总库容9853亿立方米。其中:大型水库805座,总库容7944亿立方米;中型水库4174座,总库容1197亿立方米。

堤防和水闸。截至2021年年底,全国已建成5级及以上江河堤防33.1万公里❶,累计达标堤防24.8万公里,堤防达标率为74.9%;其中,1级、2级达标堤防长度为3.8万公里,达标率为84.3%。全国已建成江河堤防保护人口6.5亿人,保护耕地4.2万千公顷。全国已建成流量为5立方米每秒及以上的水闸100321座,其中大型水闸923座。按水闸类型分,分洪闸8193座,排(退)水闸17808座,挡潮闸4955座,引水闸13796座,节制闸55569座。

---

❶ 2011年以前各年堤防长度含部分地区5级以下江河堤防长度。

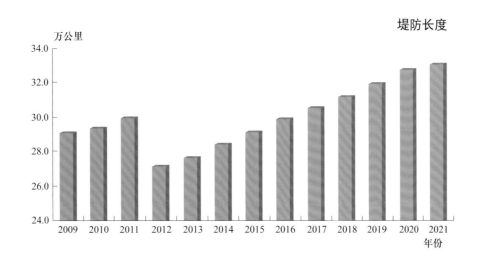

堤防长度

机电井和泵站。全国已累计建成日取水量大于等于20立方米的供水机电井或内径大于等于200毫米的灌溉机电井共522.2万眼。全国已建成各类装机流量1立方米每秒或装机功率50千瓦以上的泵站93699处，其中：大型泵站444处，中型泵站4439处，小型泵站88816处。

灌区工程。全国已建成设计灌溉面积2000亩及以上的灌区共21619处，耕地灌溉面积39727千公顷。其中：50万亩及以上灌区154处，耕地灌溉面积12209千公顷；30万~50万亩大型灌区296处，耕地灌溉面积5659千公顷。截至2021年年底，全国灌溉面积78315千公顷，耕地灌溉面积69609千公顷，占全国耕地面积的51.6%。

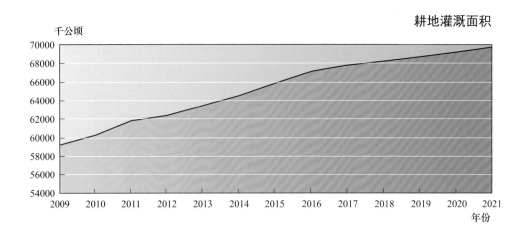

农村水电。截至 2021 年年底，全国共建成农村水电站 42785 座，装机容量 8290.3 万千瓦，占全国水电装机容量的 21.2%。全国农村水电年发电量 2241.1 亿千瓦·时，占全口径水电发电量的 16.7%。

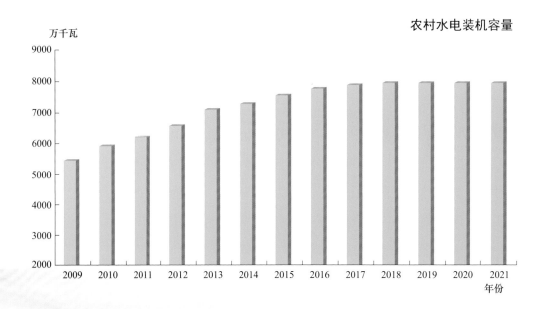

水土保持工程。全国水土流失综合治理面积达 149.6 万平方公里❶，累计封禁治理保有面积达 28.9 万平方公里。2021 年持续开展全国全覆盖的水土流失动态监测工作，全面掌握县级以上行政区、重点区域、大江大河流域的水土流失动态变化。

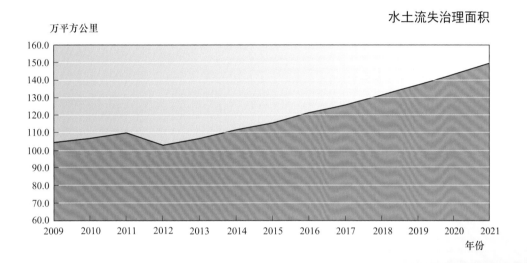

水土流失治理面积

水文站网。全国已建成各类水文测站 119491 处，包括国家基本水文站 3293 处、专用水文站 4598 处、水位站 17485 处、雨量站 53239 处、蒸发站 9 处、地下水站 26699 处、水质站 9621 处、墒情站 4487 处、实验站 60 处。其中，向县级以上水行政主管部门报送水文信息的各类水文测站 70261 处，可发布预报站 2521 处，可发布预警站 2583 处；配备在线测流系统的水文测站 2524 处，配备视频监控系统的水文测站 5331 处。基本建成中央、流域、省级和地市级共 337 个水质监测（分）中心和水质站（断面）组成的水质监测体系。

---

❶ 2012 年数据与第一次全国水利普查数据进行了衔接。

水利网信。截至 2021 年年底,全国省级以上水利部门配置累计各类服务器 9945 台(套),形成存储能力 36.26 拍字节,存储各类信息资源总量达 6.06 拍字节;县级以上水利部门累计配置各类卫星设备 3018 台(套),利用北斗卫星短文传输报汛站达 8015 个,应急通信车 64 辆,集群通信终端 3065 个,宽、窄带单通信系统 487 套,无人机 1718 架。全国省级以上水利部门各类信息采集点达 42.95 万处,其中:水文、水资源、水土保持等采集点约 20.45 万处,大中型水库安全监测采集点约 22.5 万处。

4 水资源节约利用与保护

水资源状况。 2021年，全国水资源总量29638.2亿立方米，比多年平均值偏多7.3%。全国年平均降水量❶691.6毫米，比多年平均值偏多7.4%，较上年减少2.1%。全国728座大型水库和3797座中型水库年末蓄水总量4449.1亿立方米，比年初增加17.5亿立方米。

水资源开发。 2021年，新增规模以上水利工程❷供水能力79.5亿立方米。截至2021年年底，全国水利工程供水能力达8984.2亿立方米，其中：跨县级区域供水工程631.5亿立方米，水库工程2442.5亿立方米，河湖引水工程2120.8亿立方米，河湖泵站工程1851.4亿立方米，机电井工程1383.7亿立方米，塘坝窖池工程373.5亿立方米，非常规水资源利用工程180.9亿立方米。

---

❶ 2021年全国年平均降水量依据约18000个雨量站观测资料评价。
❷ 规模以上水利工程包括：总库容大于等于10万立方米的水库、装机流量大于等于1立方米每秒或装机容量大于等于50千瓦的河湖取水泵站、过闸流量大于等于1立方米每秒的河湖引水闸、井口井壁管内径大于等于200毫米的灌溉机电井和日供水量大于等于20立方米的机电井。

水资源利用。 2021年,全国供水总量5920.2亿立方米,其中:地表水源供水量4928.1亿立方米,地下水源供水量853.8亿立方米,其他水源供水量138.3亿立方米。全国用水总量5920.2亿立方米,其中:生活用水量909.4亿立方米,工业用水量1049.6亿立方米,农业用水量3644.3亿立方米,人工生态环境补水量316.9亿立方米。与上年比较,用水总量增加107.3亿立方米,其中:生活用水量增加46.3亿立方米,工业用水量增加19.2亿立方米,农业用水量增加31.9亿立方米,人工生态环境补水量增加9.9亿立方米。

水资源节约。 全国人均综合用水量为419立方米,农田灌溉水有效利用系数0.568,万元国内生产总值(当年价)用水量51.8立方米,万元工业增加值(当年价)用水量28.2立方米。按可比价计算,万元国内生产总值用水量和万元工业增加值用水量分别比2020年下降5.8%和7.1%。全国非常规水源利用量138.3亿立方米,其中:再生水利用量117.1亿立方米,集蓄雨水利用量6.9亿立方米,淡化海水利用量2.8亿立方米,微咸水利用量3.4亿立方米,矿坑水利用量8.0亿立方米。

5 防汛抗旱

2021年，全国洪涝灾害总体偏重，洪涝灾害直接经济损失2458.9亿元（水利设施直接损失481亿元），占当年国内生产总值的0.22%。全国农作物受灾面积4760.4千公顷，绝收面积872.4千公顷，受灾5901万人次，因灾死亡512人，失踪78人，倒塌房屋15.2万间❶。河南、四川、山西、陕西、河北等省受灾较重。全国因山洪灾害造成人员死亡171人，占全部死亡人数的33.4%。

全国受旱地域分布较广，但造成的影响总体较轻，广东、浙江、福建、内蒙古等省（自治区）旱灾较重。全国农田因旱受灾面积4448千公顷，成灾面积❷2277千公顷，直接经济总损失177亿元。全国因旱累计有546万城乡人口、251万头大牲畜发生临时性饮水困难。全年完成抗旱浇地面积3998千公顷，抗旱挽回粮食损失56.3亿公斤，解决了535万城乡居民和205万头大牲畜因旱临时饮水困难。

---

❶ 2021年洪涝灾害直接经济损失、全国农作物受灾面积、绝收面积、受灾人口、因灾死亡和失踪人口、倒塌房屋数量等数据来源于应急管理部国家减灾中心。

❷ 因机构改革后职能调整，水利部不再统计发布水灾成灾面积。文中成灾数据自2019年起不再包含水灾。

全年中央下拨水利救灾资金29亿元，其中：防汛资金22.5亿元，抗旱资金6.5亿元。水利救灾资金在水旱灾害防御工作中发挥了显著作用，为保障防洪安全、供水安全提供了有力支撑。

6 水利改革与管理

节约用水管理。2021年，以县域为单元全面开展节水型社会达标建设，复核发布第4批478个节水型社会达标县（市、区、旗）。新发布3415项国家和省级用水定额。开展规划和建设节水评价项目10065个，从严叫停项目243个。推广合同节水管理，推动实施合同节水管理项目93项，吸引社会资本约1.87亿元，预计年节水量1143万立方米。加强节水技术推广应用，征集推广先进工艺、技术和装备192项。开展节水载体示范建设，新建成2429家水利行业节水型单位、262所节水型高校，遴选发布168家公共机构水效领跑者、15处灌区水效领跑者。推动计划用水，覆盖水资源超载区99.1%工业企业。将13663个用水单位纳入重点监控体系，实际监控用水总量占全国用水总量的31%。

河（湖）长制。2021年，31个省（自治区、直辖市）党委和政府主要负责同志全部担任双总河长，明确省、市、县、乡级河长湖长30万名，村级河长湖长（含巡河员、护河员）90多万名，实现河湖管护责任全覆盖。全国省、市、县、乡级河长湖长累计巡查河湖594万人次。深入推进河湖"清四乱"常态化规范化，全国共清理整治"四乱"

问题 2.9 万个，拆除侵占河湖违建 810 多万平方米，清理非法占用岸线 7000 多公里，清除河道内垃圾 810 多万吨，清除河道非法采砂点 2700 多个，打击非法采砂船只 1000 多艘，河湖面貌持续改善。全面完成长江干流岸线利用项目清理整治任务，2441 个长江干流违法违规岸线利用项目全部完成清理整治，腾退长江岸线 162 公里，完成复绿 1200 多万平方米；按照中央关于长江"十年禁渔"决策部署，组织开展长江非法矮围专项整治，对长江干流和洞庭湖、鄱阳湖排查发现的 63 处非法矮围进行清理整治，拆除非法围堤 59 公里，恢复水域面积 6.8 万亩；与公安部、交通运输部、工业和信息化部、市场监管总局联合开展长江河道采砂综合整治和采砂船舶专项治理，查处非法采砂案件 1867 起，查获非法采砂船 185 艘（其中隐形采砂船 27 艘），向公安机关移送非法采砂案件 104 件。强化监督检查，对 3234 条河流（5192 个河段）、991 个湖泊（1124 个湖片）开展暗访检查，覆盖 31 个省（自治区、直辖市）所有设区市和流域面积在 1000 平方公里以上河流、水域面积 1 平方公里以上湖泊（除无人区外）；对长江、黄河、大运河等重点流域的 8 个省份开展进驻式检查。

水资源管理。 加快推进水资源管控指标制定，累计批复 63 条跨省江河水量分配方案；组织完成 82 个跨省重要河湖的生态流量保障目标制订，指导各省完成 134 个重点河湖生态流量保障目标制订；13 个省（自治区、直辖市）印发了地下水取水总量控制、水位控制"双控"指标；明确了"十四五"全国用水总量控制目标，并分解到各省区；扎实推进取用水管理专项整治行动，全面完成取水口核查登记，基本摸清了超过 580 万个取水口的合规性和取水监测计量现状，并开展整改。

组织完成 5 个国家农业高新技术产业示范区规划水资源论证审查。对黄河流域 13 个地表水超载地市、62 个地下水超采县暂停新增取水许可。取水许可电子证照实现全面应用，累计发放电子证照 52 万余张，基本完成存量证照电子化转换。推进南水北调受水区地下水压采，累计压采地下水 30.17 亿立方米。完成"十三五"最严格水资源管理制度考核，考核结果报国务院审定后公布，将考核结果报中组部，作为党政领导干部综合考核评价的重要依据。中国水权交易所 2021 年完成水权交易 1511 单，交易水量 3.08 亿立方米。

流域水资源统一调度。 2021年，水利部印发《水资源调度管理办法》，对水资源调度权限、依据、组织、实施、监测、监督管理、责任追究等方面问题予以规定。2020—2021调水年度，南水北调东线一期工程向山东省调水6.74亿立方米，中线一期工程向北京、天津、河北、河南四省（直辖市）调水共计90.54亿立方米，受水区供水安全保障能力显著提升。加快推进江河流域水资源统一调度，实施汉江、嘉陵江、乌江等31条跨省江河统一调度，珠江流域通过实施17次枯水期水量调度全面保障了澳门、珠海等地供水安全。强化重点流域生态统一调度，黄河干流开展生态调度实现连续22年不断流。实施华北地区河湖常态化补水和夏季集中贯通补水，2021年累计向华北地区22条河湖补水84.68亿立方米，永定河、潮白河、滹沱河、大清河等多年断流河道全线贯通。乌梁素海通过补水5.98亿立方米，生态环境持续向好，黑河下游东居延海连续17年不干涸，永定河实现自1996年以来首次全线通水入海。

运行管理。 截至2021年年底，累计批准国家级水利风景区902个，其中：水库型379个，自然河湖型202个，城市河湖型204个，湿地型47个，灌区型32个，水土保持型38个。

水价改革。 配合国家发展改革委出台《"十四五"时期深化价格机制改革行动方案》，明确健全有利于促进水资源节约和水利工程良性运行、与投融资体制相适应的水利工程水价形成机制。推动修订《水利工程供水价格管理办法》《水利工程供水定价成本监审办法》，进一步完善水价形成机制和动态调整机制。截至 2021 年年底，累计实施农业水价综合改革面积 6 亿亩，其中 2021 年新增农业水价综合改革面积 1.6 亿亩。

水利规划和前期工作。2021 年，国家层面审批（含印发审查意见）水利规划 38 项。经国务院同意，会同国家发展改革委联合印发"十四五"水安全保障规划，同时编制和印发了一系列配套规划或实施方案，形成"1+N"的规划体系。会同国家发展改革委组织编制国家水网建设规划纲要，上报国务院。扎实推进国家重大战略水利规划工作，全力抓好长江三角洲区域一体化发展、黄河流域生态保护和高质量发展、成渝地区双城经济圈发展、太湖流域水环境综合治理等国家重大区域发展战略水利各项任务落实。加快推进重点流域和主要支流综合规划审批，批复岷江、韩江、拉林河等流域综合规划，加快其他支流规划编制进度。组织开展七大流域防洪规划修编，编制完成七大流域防洪规划修编任务书。2021 年，国家发展改革委批复工程可行性研究报告 9 项，其中防洪工程 5 项、水资源配置工程 4 项，投资规模 407.03 亿元。水利部批复初步设计 8 项，其中防洪工程 5 项、水资源配置工程 3 项，投资规模 355.37 亿元。

水土保持管理。2021 年，水利部印发《水土保持"十四五"实施方案》，出台《水利部关于进一步推动水土保持工程建设以奖代补的指导意见》《水利部办公厅关于加强水利建设项目水土保持工作的通知》。全国共审批生产建设项目水土保持方案 11.19 万个，涉及水土流失防治责任范围 2.43 万平方公里；4.77 万个生产建设项目完成水土保持设施自主验收报备。开展覆盖全国范围（不含港澳台）的生产建设项目人为水土流失遥感监管，通过卫星遥感解译组织现场复核，共认定并查处"未批先建""未批先弃"等违法违规项目 2.41 万个。开展国家水土保持示范创建，共评定 90 个示范；选取江西赣州、陕西延安、福建长汀、山西右玉、黑龙江拜泉等 5 个市县开展全国水土保持高质量发展先行区建设。

农村水电管理。截至 2021 年年底，25 个省份累计创建绿色小水电站 870 座。积极推进农村水电站安全生产标准化建设，全国累计建成安全生产标准化电站 3963 座，其中一级电站 99 座、二级电站 1498 座、三级电站 2176 座。

水利移民。2021 年搬迁人口 20.09 万人，其中农村移民搬迁 18.51 万人，城集镇移民搬迁 1.58 万人，生产安置 22.05 万人。国家核定 2020 年度新增大中型水库农村移民后期扶持人数 15.02 万人。

水利监督。2021 年，组织开展了"实行最严格水资源管理制度考核""水旱灾害防御""水利工程建设及运行管理""水土保持及水行政执法履职"等中央计划事项和备案事项的监督检查，覆盖防汛、水资源管理、农村饮水安全等 32 项内容。全年共派出检查组 2321 个、10221 人次，检查项目 21700 个，对 387 家责任单位实施了"约谈"及以上责任追究。扎实开展安全生产专项整治三年行动和隐患排查整治专项行动，隐患排查整改率99.6%，全行业制定有关制度2550个。加强工程建设、运行重点领域及关键环节的安全巡查和监督检查，组织印发"一省一单"整改意见46份，对5家责任单位实施"约谈"。积极推进安全生产标准化工作，全年131家单位通过标准化一级达标创建评审。深入推进"安全监管+信息化"，2021年第四季度全国平均安全风险度为14.26，同比2020年降低45.8%，行业整体安全风险可控。推进重大水利工程质量与安全巡查，累计派出75人次，发现质量安全问题162个，印发"一省一单"整改意见通知13份。加大对流域防洪工程和国家水网重大工程的稽察力度，全年开展稽察7批次，共派出稽察组73个、675人次，涉及重大水利工程、水库除险加固等项目127个，印发"一省一单"整改意见通知58份，对86家责任单位实施"约谈"及以上责任追究。推动省级水行政主管部门开展自主稽察150余批次，共派出稽察组460余个，近3900人次，涉及项目近1070个。

依法行政。2021年，出台水法律1件、行政法规1件，完成法律法规起草3件。2021年全国立案查处水事违法案件1.98万件，结案1.86万件，结案率93.9%；水利部共办结行政复议案件12件，办理行政应诉20件。

行政许可。2021年，水利部（包括部机关和各流域机构）共受理行政审批事项2561件，办结1727件。其中：水工程建设规划同意书审核39项，不同行政区域边界水工程批准3件，水利基建项目初步设计文件审批8件，取水许可154件，非防洪建设项目洪水影响评价报告审批31件，河道管理范围内建设项目工程建设方案审批396件，生产建设项目水土保持方案审批64件，国家基本水文测站设立和调整审批3件，国家基本水文测站上下游建设影响水文监测工程的审批113件，水利工程建设监理单位资质认定（新申请、增项、晋升、延续）1368件，水利工程质量检测单位甲级资质认定（新申请、增项、晋升、延续）415件。

水利科技。2021年，国家立项安排4.04亿元资金用于水利科技项目，其中：组织承担国家重点研发计划"长江黄河等重点流域水资源与水环境综合治理""重大自然灾害防控与公共安全"等涉水重点专项项目共9项，合计23855万元；国家自然科学基金长江、黄河水科学研究联合基金项目48项，合计1.5亿元；水利技术示范项目58项，合计1483.86万元，完成55项结题备案。水利科技成果获国家科技进步奖二等奖1项。截至2021年年底，水利系统共有国家和部级重点实验室

17个（含筹建中部级重点实验室5个），国家和部级工程技术研究中心15个。落实中央财政公益性科研院所基本科研业务费9401万元。发布水利技术标准34项，在编145项，水利行业现行有效标准达802项。

国际合作。2021年，共签署水利国际合作协议1份，在华举办多边、双边高层圆桌会议或技术交流研讨会35次，线上参加涉水国际组织机制性会议50场。利用世界银行、亚洲开发银行、全球环境基金贷、增款开展的8个项目进展顺利，中瑞、中丹、中法、中芬合作项目和国际科技合作项目稳步开展。落实澜湄合作第三次领导人会议成果，成功举办第二届澜湄水资源合作论坛；组织实施亚洲合作资金项目9个，落实经费1963万元；应急协调俄罗斯与我方加强跨界水水文信息和支流重要工程调度信息共享，深化防洪领域合作；圆满完成对周边国家72个水文站国际报汛工作。

7 水利行业状况

水利单位。 截至 2021 年年底，从事水利活动的各类县级及以上独立核算的法人单位 20413 个，从业人员 86.5 万人。其中：机关单位 2704 个，从业人员 12.5 万人，比去年增加 0.8%；事业单位 13729 个，从业人员 45.9 万人，比去年减少 5.0%；企业 3332 个，从业人员 27.3 万人，比去年减少 5.2%；社团及其他组织 648 个，从业人员 0.6 万人，比去年增加 50%。

职工与工资。 全国水利系统从业人员 77.9 万人，比上年减少 3.72%。其中，全国水利系统在岗职工 74.8 万人，比上年减少 3.81%。在岗职工中，部直属单位在岗职工 6 万人，比上年减少 9.76%；地方水利系统在岗职工 68.8 万人，比上年减少 3.25%。全国水利系统在岗职工工资总额为 818.7 亿元，年平均工资为 11 万元。

职工与工资情况

	2011年	2012年	2013年	2014年	2015年	2016年	2017年	2018年	2019年	2020年	2021年
在岗职工人数/万人	102.5	103.4	100.5	97.1	94.7	92.5	90.4	87.9	82.7	77.8	74.8
其中：部直属单位/万人	7.5	7.4	7.0	6.7	6.6	6.4	6.4	6.6	6.6	6.7	6.0
地方水利系统/万人	95.0	96.0	93.5	90.4	88.1	86.1	84.0	81.3	76.0	71.1	68.8
在岗职工工资/亿元	351.4	389.1	415.3	451.4	529.4	640.5	739.1	802.7	787.6	790.9	818.7
年平均工资/(元/人)	34283	37692	41453	46569	55870	69377	83534	91307	95385	102000	110000

全国水利发展主要指标（2016—2021年）

指标名称	单位	2016年	2017年	2018年	2019年	2020年	2021年
1. 灌溉面积	千公顷	73177	73946	74542	75034	75687	78315
2. 耕地灌溉面积	千公顷	67141	67816	68272	68679	69161	69609
其中：本年新增	千公顷	1561	1070	828	780	870	1114
3. 节水灌溉面积	千公顷	32847	34319	36135	37059	37796	
其中：高效节水灌溉面积	千公顷	19405	20551	21903	22640	23190	
4. 万亩以上灌区	处	7806	7839	7881	7884	7713	7326
其中：30万亩以上	处	458	458	461	460	454	450
万亩以上灌区耕地灌溉面积	千公顷	33045	33262	33324	33501	33638	35499
其中：30万亩以上	千公顷	17765	17840	17799	17994	17822	17868
5. 农村自来水普及率	%	79	80	81	82	83	84
6. 除涝面积	千公顷	23067	23824	24262	24530	24586	24480
7. 水土流失治理面积	万平方公里	120.4	125.8	131.5	137.3	143.1	149.6
其中：新增	万平方公里	5.6	5.9	6.4	6.7	6.4	6.8
8. 水库	座	98460	98795	98822	98112	98566	97036
其中：大型水库	座	720	732	736	744	774	805
中型水库	座	3890	3934	3954	3978	4098	4174
水库总库容	亿立方米	8967	9035	8953	8983	9306	9853
其中：大型水库	亿立方米	7166	7210	7117	7150	7410	7944
中型水库	亿立方米	1096	1117	1126	1127	1179	1197
9. 全年水利工程总供水量	亿立方米	6040	6043	6016	6021	5813	5920
10. 堤防长度	万公里	29.9	30.6	31.2	32.0	32.8	33.1
保护耕地	千公顷	41087	40946	41409	41903	42168	42192
堤防保护人口	万人	59468	60557	62837	67204	64591	65193
11. 水闸总计	座	105283	103878	104403	103575	103474	100321
其中：大型水闸	座	892	892	897	892	914	923

续表

指标名称	单位	2016年	2017年	2018年	2019年	2020年	2021年
12. 年末全国水电装机容量	万千瓦	33153	34168	35226	35564	36972	39184
全年发电量	亿千瓦·时	11815	11967	12329	12991	13540	13419
13. 农村水电装机容量	万千瓦	7791	7927	8044	8144	8134	8290
全年发电量	亿千瓦时	2682	2477	2346	2533	2424	2241
14. 当年完成水利建设投资	亿元	6099.6	7132.4	6602.6	6711.7	8181.7	7576.0
按投资来源分：							
（1）中央政府投资	亿元	1679.2	1757.1	1752.7	1751.1	1786.9	1708.6
（2）地方政府投资	亿元	2898.2	3578.2	3259.6	3487.9	4847.8	4236.8
（3）国内贷款	亿元	879.6	925.8	752.5	636.3	614.0	698.9
（4）利用外资	亿元	7.0	8.0	4.9	5.7	10.7	8.1
（5）企业和私人投资	亿元	424.7	600.8	565.1	588.0	690.4	718.2
（6）债券	亿元	3.8	26.5	41.6	10.0	87.2	104.3
（7）其他投资	亿元	207.1	235.9	226.3	232.8	144.9	101.1
按投资用途分：							
（1）防洪工程	亿元	2077.0	2438.8	2175.4	2289.8	2801.8	2497.0
（2）水资源工程	亿元	2585.2	2704.9	2550.0	2448.3	3076.7	2866.4
（3）水土保持及生态建设	亿元	403.7	682.6	741.4	913.4	1220.9	1123.6
（4）水电工程	亿元	166.6	145.8	121.0	106.7	92.4	78.8
（5）行业能力建设	亿元	56.9	31.5	47.0	63.4	85.2	79.9
（6）前期工作	亿元	174.0	181.2	132.0	132.7	157.3	136.5
（7）其他	亿元	636.2	947.5	835.8	757.4	747.3	793.8

说明：1. 本公报不包括香港特别行政区、澳门特别行政区及台湾省的数据。
　　　2. 水利发展主要指标分别于2012年、2013年与第一次全国水利普查数据进行了衔接。
　　　3. 农村水电的统计口径为单站装机容量5万千瓦及以下的水电站。

2021 STATISTIC BULLETIN ON CHINA WATER ACTIVITIES

Ministry of Water Resources, P. R. China

The year of 2021 is the 100th anniversary of the founding of the Communist Party of China (CPC) and a milestone year in the history of the CPC and the country. Over the past year, under the strong leadership of the Central Committee of the Party and the State Council, water resources departments at all levels have been implementing the new development concept in a complete, accurate and comprehensive way, focusing on top priority and agenda of the country, implementing the guidance of General Secretary Xi Jinping to "prioritize water conservation, balance development with spatial distribution of water resources, practice systematic governance and achieve government-market synergy", as well as instructions in his keynote speeches on water governance. The 14th Five-Year Plan has a good beginning as we snatch genuine priorities and endeavor in pragmatic way, overcome difficulties and forge ahead, and take a strong step to advance high-quality water development in the new stage.

I. Investment in Fixed Assets

In 2021, the total investment in water projects amounted to 757.60 billion Yuan, among which, 585.13 billion Yuan was being allocated for construction projects, accounting for 77.2% of the national total. 33.01 billion Yuan for installation, accounting for 4.4% of the national total. 20.36 billion Yuan for expenditure on purchases of machinery, electric equipment and instruments, accounting for 2.7% of the national total, and 119.10 billion Yuan for other purposes, including compensation for resettlement and land acquisition, accounting for 15.7% of the national total.

	2014 /billion Yuan	2015 /billion Yuan	2016 /billion Yuan	2017 /billion Yuan	2018 /billion Yuan	2019 /billion Yuan	2020 /billion Yuan	2021 /billion Yuan
Total completed investment	408.31	545.22	609.96	713.24	660.26	671.17	818.17	757.60
Construction project	308.64	415.08	442.20	506.97	487.72	498.79	601.49	585.13
Installation project	18.50	22.88	25.45	26.58	28.09	24.31	31.97	33.01
Purchase of machinery, equipment and instruments	20.61	19.87	17.28	21.17	21.44	22.11	25.00	20.36
Others (including compensation of resettlement and land acquisition)	60.56	87.39	125.03	158.52	123.01	125.97	159.71	119.10

In the total completed investment, 249.70 billion Yuan was allocated to the construction of flood control projects, accounting for 33.0% of the national total; 286.64 billion Yuan was for the construction of water resources projects, accounting for 37.8% of the national total; 112.36 billion Yuan was for soil and water conservation and ecological restoration, accounting for 14.8% of the national total; and 108.89 billion Yuan for specific projects of hydropower development and capacity building, accounting for 14.4% of the national total.

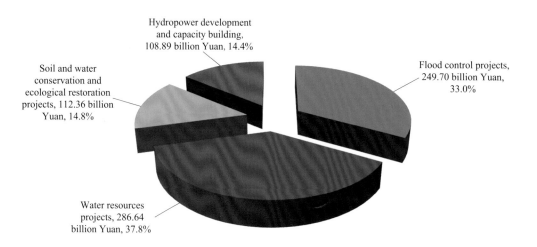

Completed investment of projects in 2021

The competed investment for seven major river basins reached 611.24 billion Yuan, of which 146.36 billion Yuan was invested in river basins in the southeast, southwest and northwest of China, while investments in river basins in east, central, west and northeast China were 316.60 billion Yuan, 185.56 billion Yuan, 232.04 billion Yuan and 23.40 billion Yuan, respectively.

Of the total competed investment, the Central Government contributed 6.78 billion Yuan, and local governments contributed 750.82 billion Yuan. Investments for large and medium-sized projects were 169.01 billion Yuan; and for small and other projects were 588.59 billion Yuan. Investments for new projects and rehabilitation and expansion projects were 574.60 billion Yuan and 183 billion Yuan.

The newly-added fixed asset of the year in water project construction totaled 401.78 billion Yuan. By the end of 2021, the accumulated investment in projects under construction was 1,835.01 billion Yuan, with the completion rate reaching 62.2%. New fixed assets totaled 988.84 billion Yuan and the rate of investment transferred into fixed assets was 53.0%, an increase of 3.6 percentage point over the previous year.

A total of 31,614 water projects were under construction in 2021, with a total investment of 2,950.21 billion Yuan, a decrease of 7% over the previous year, among which, 15,936 projects were funded by the Central Government, a decrease of 2.3% over the previous year, occupying 1,255.32 billion Yuan, down 5.7% over the previous year. There were 20,900 projects launched in 2021, a decrease of 7.2%, with an increase of investment totaled 666.43 billion Yuan, a decrease of 16% over the previous year. The completed civil works of earth, stone and concrete structures were 2.68 billion m^3, 380 million m^3, and 240 million m^3, respectively. By the end of 2021, the rate of completed quantity of earthwork, stonework, and concrete of under-construction projects were 97.5%, 98.9% and 89.6%, respectively.

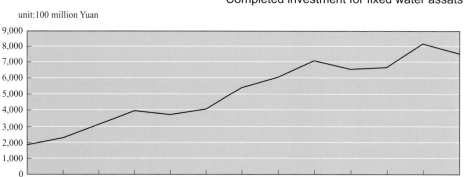

II. Key Water Projects Construction

Harness of large rivers and lakes. In 2021, there were 4,903 river harness projects under construction, including 643 flood control dykes and embankments, 713 projects for large river and main tributary control, and 2,928 medium-sized and small river control works, and 619 flood diversion and storage area construction or other projects. By the end of 2021, the accumulated investment in projects under construction was 391.67 billion Yuan, with a completion rate of 67%. 9 projects for river regime control and river course training and restoration in the middle and lower reaches of the Yangtze River were completed for benefit generation. Flood control works in the lower reaches of the Yellow River were all completed and 2 projects have started to generate benefits. Out of the 38 Huaihe River improvement projects, 34 started construction, among which 16 were put into operation for benefit generation. 5 projects of improvement work for Dongting and Poyang lakes were opened to construction and three of which were completed for benefit generation. 12 projects for the Comprehensive Improvement of Water Environment of Taihu Lake began construction, 8 of which completed construction with benefit generation.

Reservoir and water control projects. In 2021, there were 1,275 reservoir projects under construction. By the end of 2021, the completed investment of under-construction projects reached 353.68 billion Yuan, with a completion rate of 67.5%. Construction commenced for 19 medium-sized reservoir projects, including Yuanwan Reservoir in Henan, Guanyin Reservoir in Guizhou and Xiwei Reservoir. Dongtaizi Reservoir in Inner Mongolia, Dongzhuang Multipurpose Dam Project in Shaanxi and Maiwan Multipurpose Dam Project in Hainan realized river damming. Reservoirs of Jiangxiang in Anhui, Huangjiawan Reservoir in Guizhou, Dehou Reservoir in Yunnan and Sifangjing Multipurpose Dam Project in Jiangxi completed impoundment.

Planning, implementation and management of the follow-up work of the Three Gorges Project. In 2021, the central government allocated 11.306 billion Yuan to the Construction Fund for National Major Water Resources Project (the follow-up work of the Three Gorges Project), an increase of 35.16% over 2020, including 10.994 billion Yuan of local transfer payments and 312 million Yuan for planning, implementation and management of the follow-up work of the Three Gorges Project. The investment completed in 2021 was 9.229 billion Yuan, accounting for 81.63% of the budget. In the year, 377 projects were implemented in the Three Gorges Reservoir Area and main affected areas in the middle and lower reaches of the Yangtze River to ensure wealth of resettlement population and promote economic and social development in the reservoir area. A total of 75,000 resettled people were directly benefited from industrial development and employment arrangement plans. Subsidies were allocated to 2,857 children for higher vocational education, and 6,215 people in the reservoir area for training and employment. There are 379,600 resettled people benefited from urban assistance projects. The number of people benefited from rural assistance project reached 321,900. There were 120 ecological environment restoration and protection schemes implemented, with a total length of 140.18 km for comprehensive improvement of the reservoir bank and 31 tributaries. There were 120 geological hazard prevention and control projects implemented, to eliminate 66 geological hazard hazards, protect 2,190 people, provide monitoring and warning for 3,091 high cut slopes, and monitoring and protection for 460,400 people under the risks of high cut slope. A total of 31 projects were conducted to reduce impact of operation of the Three Gorges Reservoir on the key affected areas in the middle and lower reaches of the Yangtze River, and protect 90.08 km of bank in the affected areas of the middle and lower reaches. The implementation of 31 capacity building for comprehensive management and benefit boosting projects provided a solid base for high-quality development in the implementation of the follow-up work of the Three Gorges Project in the new stage.

Water allocation projects. In 2021, investment for water allocation projects under construction reached 691.19 billion Yuan and completed investment accumulated to 433.14 billion Yuan, accounting for 62.7% of the total. The second phase of water diversion from the Hanshui River to the Weihe River in Shaanxi and water supply to northwest of Hainan commenced construction. Progressed sped up for the projects of water diversion from the Yangtze River to the Huaihe River, water diversion in central Yunnan and water allocation in the Pearl River Delta.

Rural water supply, irrigation and drainage. In 2021, the completed investment for water supply in rural areas reached 52.5 billion Yuan that enhance the guarantee rate of water supply to 42.63 million rural population. The Central Government allocated 7.54 billion Yuan for continuous construction of large irrigation and drainage systems construction and counterpart facilities for modernization, and 7.0 billion Yuan for continuous construction of medium-sized irrigation districts and rehabilitation for water saving. The newly-added irrigated area was 1,114,000 ha. By the end of 2021, the rate of access to tap water reached 84%.

Rural hydropower and electrification. In 2021, the completed investment of rural hydropower station construction nationwide amounted to 3.32 billion Yuan, adding 61 new hydropower stations, with a total installed capacity of 312,000 kilowatts.

Soil and water conservation. In 2021, a total of 320.57 billion Yuan was allocated to the under-constructed projects for soil and water conservation and ecological restoration, with an accumulated investment of 174.16 billion Yuan. The newly-added areas for comprehensive control of soil erosion reached 68,000 km^2, of which the areas under the National Major Project for Soil Conservation was 12,800 km^2. Up to 556 silt-retention dams on Loess Plateau at high risk were strengthened and rehabilitated. The area of slope cultivated land is 860,000 mu, and 684 warping dams (Stand retaining dams) are nealy built.

Capacity building. The completed investment for capacity building in 2021 was 6.01 billion Yuan, of which 510 million Yuan was spent on communication equipment for flood control, 2.38 billion Yuan on hydrological construction, 180 million Yuan for scientific research and education facilities and 2.94 billion Yuan for others.

III. Key Water Structures and Facilities

Reservoirs and water complexes. The number of completed reservoirs in China reached 97,036, with a total storage capacity of 985.3 billion m^3. Of which 805 reservoirs are large reservoirs, with a total capacity of 794.4 billion m^3 and 4,174 reservoirs are medium-sized with a total capacity of 119.7 billion m^3.

Embankments and water gates. By the end of 2021, the completed river dykes and embankments at Grade-V or above had a total length of 331,000 km❶. The accumulated length of dykes and embankments that met the standard reached 248,000 km, accounting for 74.9% of the total. Among which, Grade-I and Grade-II dykes and embankments were up to the standard reached 38,000 km, or 84.3% of the total. All completed dykes and embankments nationwide can protect 650 million people and 42 million ha of cultivated land. The number of completed water gates with a flow of 5 m^3/s increased to 100,321, of which 923 were large water gates. By type, there were 8,193 flood diversion sluices, 17,808 drainage/return water sluices, 4,955 tidal barrages, 13,796 water diversion intakes and 55,569 controlling gates.

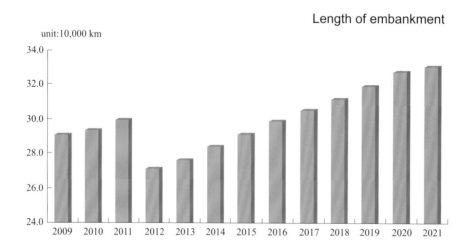

❶ The length of river dykes in each year before 2011 included the length of river dykes below Grade-V in some areas.

Tube wells and pumping stations. Accumulatively, a total of 5.222 million tube wells, with a daily water abstraction capacity equal to or larger than 20 m^3 or an inner diameter equal to or larger than 200 mm, had been completed for water supply in the whole country. A total of 93,699 pumping stations with a flow of 1 m^3/s or an installed voltage above 50 kW were put into operation, including 444 large, 4,439 medium and 88,816 small pumping stations.

Irrigation systems. The irrigation districts with a designed area of 2,000 mu or above were 21,619 in total, covering 39.727 million ha of irrigated farmland. Of which 154 irrigation districts had an irrigated area of 500,000 mu or above, and their total irrigated area reached 12.209 million ha. The irrigation districts with an area from 300,000 to 500,000 mu were 296, covering 5.659 million ha of irrigated farmland. By the end of 2021, the total irrigated area amounted to 78.315 million ha. The irrigated area of cultivated land reached 69.609 million ha that accounted for 51.6% of the total in China.

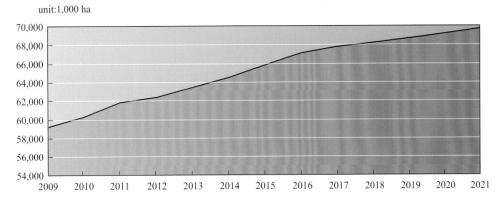

Irrigated Area of cultivated land

Rural hydropower and electrification. By the end of 2021, hydropower stations built in rural areas totaled 42,785, with an installed capacity of 82.903 million kW, accounting for 21.2% of the national total. The annual power generation by these hydropower stations reached 224.11 billion kW·h, accounting for 16.7% of the national total.

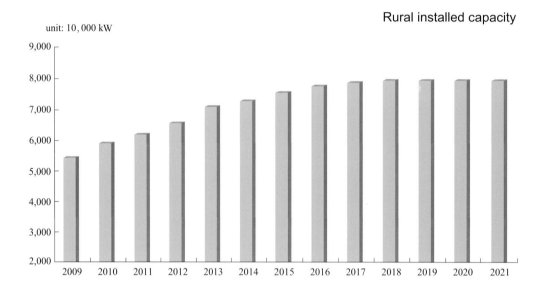

Soil and water conservation. By the end of 2021, the restored eroded areas reached 1.496 million km^2;❶ and the forbidden area for ecological restoration accumulated to 289,000 km^2. Dynamic monitoring for soil and water loss had been continued in 2021 in all administrative areas above county level, key areas, and major river basins in the country, to gain a comprehensive understanding of dynamic changes.

---

❶ The data in 2012 were linked to the data of the First National Water Resources Survey.

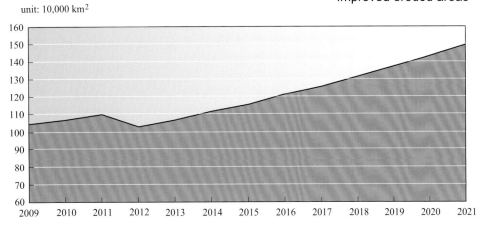

Improved eroded areas

Hydrological station networks. In 2021, the number of hydrological stations of all kinds totaled 119,491 in the whole country, including 3,293 national basic hydrologic stations, 4,598 special hydrologic stations, 17,485 gauging stations, 53,239 precipitation stations, 9 evaporation stations, 26,699 groundwater monitoring stations, 9,621 water quality stations, 4,487 soil moisture monitoring stations and 60 experimental stations. Among them, 70,261 stations of various kinds can provide hydrological information to water administration authorities at and above county level; 2,521 stations can provide forecasting and 2,583 may issue early warnings; 2,524 were equipped with online flow measurement and 5,331 were equipped with video monitors. A water quality monitoring system, including 337 monitoring centers and sub-centers as well as water quality stations (sections) at central, basin, provincial and local levels, had been set up.

Water networks and information systems. By the end of 2021, the water resources departments and authorities at and above provincial level were equipped with 9,945 servers of varied kinds, forming a total storage capacity of 36.26PB, and keeping 6.06PB of data and information. The water resources departments and authorities at and above county level had equipped with 3,018 sets of various kinds of satellite equipment, 8,015 flood forecasting stations for short message transmission from the Beidou Satellites, 64 vehicles for emergency communication, 3,065 cluster communication terminals, 487 narrowband and broadband communication systems, and 1,718 unmanned aerial vehicles (UAV). A total of 429,500 information gathering points were available for water resources departments and authorities at and above county level, including 204,500 points for collecting data of hydrology, water resources and soil and water conservation and 225,000 points for safety monitoring at large and medium-sized reservoirs.

IV. Water Resources Conservation, Utilization and Protection

Water resources conditions. The total national water resources in 2021 was 2,963.82 billion m^3, approximately 7.3% more than the normal years. The mean annual precipitation ❶ was 691.6 mm, 7.4% more than normal years and 2.1% less than the previous year. The total storage of 728 large and 3,797 medium-sized reservoirs were 444.91 billion m^3, adding 1.75 billion m^3 of water to that at the beginning of the year.

Water resources development. In 2021, the newly-increased water supply capacity by water facilities above designated size❷ was 7.95 billion m^3. By the end of 2021, the total water supply capacity of China reached 898.42 billion m^3, among which 63.15 billion m^3 was from water supply system at the county level, 244.25 billion m^3 from reservoirs, 212.08 billion m^3 from water diversion from rivers and lakes, 185.14 billion m^3 from pumping stations along rivers and lakes, 138.37

---

❶ Average precipitation of 2021 was based on data from approximately 18,000 stations.

❷ Water projects above designated size include reservoirs with a total capacity of 100,000 m^3 or higher, pump stations with an installed flow at or above 1 m^3/s or an installed capacity at or above 50 kW, water diversion gates with a flow at or above 1 m^3/s, electric irrigation wells 200 mm or larger in inner diameter or with a water supply capacity at or above 20 m^3 per day.

billion m³ from electro-mechanical wells, 37.35 billion m³ from ponds, weirs and cellars, and 18.09 billion m³ from unconventional water sources.

Water resources utilization. In 2021, the total quantity of water supply amounted to 592.02 billion m³, including 492.81 billion m³ from surface water, 85.38 billion m³ from groundwater and 13.83 billion m³ from other sources. The total water consumption amounted to 592.02 billion m³, among which domestic water use amounted to 90.94 billion m³, industrial water use totaled 104.96 billion m³, agricultural water use was 364.43 billion m³, artificial recharge for environmental and ecological use 31.69 billion m³. Comparing to the previous year, the total water consumption increased by 10.73 billion m³. Among which, domestic water use increased by 4.63 million m³, agricultural water use increased by 3.19 billion m³, industrial water use increased by 1.92 billion m³, and artificial recharge for environmental and ecological purposes increased by 0.99 billion m³.

Water Conservation Water consumption per capita in 2021 was 419 m³ in average. The coefficient of effective irrigated water use was 0.568. Water use per 10,000 Yuan of GDP (at comparable price of the same year) was 51.8 m³ and that per 10,000 Yuan of industrial value added (at comparable price of the same year) was 28.2 m³. Based on estimation at comparable prices, water uses per 10,000 Yuan of GDP and per 10,000 Yuan of industrial value added decreased by 5.8% and 7.1% over the previous year respectively. Water consumption of unconventional water sources reached 13.83 billion m³, among which 11.71 billion m³ from reclaimed water, 0.69 billion m³ from collected or stored rainwater, 0.28 billion m³ from desalinated water, 0.34 billion m³ from blackish water and 0.8 billion m³ from treated mine water.

V. Flood Control and Drought Relief

In 2021, the loss caused by flood and waterlogging is relatively severe with a direct economic loss of 245. 89 billion Yuan (including 48. 1 billion Yuan of direct losses as result of water facility damage), accounting for 0. 22% of GDP in the same year. A total of 4, 760, 400 ha of cultivated land were affected by floods, including 872, 400 ha non-harvest farmland, affected population of 59. 01 million, 512 dead and 78 missing and 152, 000 collapsed houses❶. Provinces suffered heavily from severe flooding included Henan, Sichuan, Shanxi, Shaanxi and Hebei provinces. The death toll from mountain floods was 171 that accounted to 33. 4% of the national total.

Drought was widespread but not severe on the whole. The seriously affected provinces and autonomous regions were Guangdong, Zhejiang, Fujian and Inner Mongolia. The affected area of farmland was 4, 448, 000 ha and areas with no harvest❷ reached 2, 277, 000 ha, with a total of 17. 7 billion Yuan of direct economic loss. A total of 5. 46 million urban and rural residents and 2. 51 million man-feed big animals and livestock suffered from temporary drinking water shortage. Up to 3, 998, 000 ha land were irrigated against droughts that retrieved a grain loss of 5. 63 kg. Drinking water was provided to 5. 35 million rural and urban population and 2. 05 million big animals and livestock in order to alleviate temporary water shortage.

In 2021, the Central Government allocated a total of 2. 9 billion Yuan for water-related disaster mitigation, including 2. 25 billion Yuan for flood defense and 650 million Yuan for drought relief. The funds for disaster relief have played a remarkable role in flood control and drought relief by providing strong support for safeguarding flood control and security of water supply.

---

❶ The data of direct economic losses caused by floods in 2021, the affected area of crops in China, the area of no harvest, the affected population, the number of dead and missing persons due to disasters, and the number of collapsed houses come from the National Disaster Reduction Center of Ministry of Emergency Management.

❷ Due to the adjustment of functions after the institutional reform, the Ministry of Water Resources no longer publishes statistics on flood-stricken areas. The disaster data in the figure no longer includes floods since 2019.

VI. Water Management and Reform

Water conservation management. In 2021, the construction of up-standard water-saving society at the county level was initiated with approval of 478 counties, municipalities, districts and banners in four batches. There were 3,415 national and provincial water quotas issued. The quotas for agriculture water use have covered 88% of harvested area of grain and 85% of oil crops, and quotas for service sector and industrial water use have covered 90% and 80% of the total consumption respectively. Water-saving assessment was conducted to 10,065 planned and constructed projects and 243 of which were failed and canceled. Contract-based water conservation and management was extended with 93 projects implemented, which attracted more than 187 million Yuan of social capital and saved 11.43 million m^3 of water. Extension and application of water-saving technologies was highlighted by collecting and popularizing 192 advanced technologies and equipment. In order to demonstrate results of water conservation, there were 2,429 institutions and organizations in the water sector included in the list of water-saving organization, 262 colleges and universities. There were 168 public institutions and 15 irrigation districts were selected and issued in list of leaders of water efficiency. Planed water use was promoted that covered 99.1% industrial enterprises in water overexploited areas. There were 13,663 water users fit into the key monitoring system, and the amount of water use under actual monitoring accounted to 31% of the national total.

River (lake) chief system. In 2021, all of the 31 provinces, autonomous regions and municipalities had CPC and government leaders serving as river chiefs, with more than 300,000 river (lake) chiefs named at provincial, city, county and township levels and more than 900,000 river (lake) chiefs (including those for river patrol and protection) named at village level, so as to realize full coverage of river/lake management and maintenance. The river (lake) chiefs at provincial, city, county and township levels made 5.94 million rounds of inspections on rivers and lakes. The action of punishing "misappropriation, illegal sand excavation, disposing of wastes and building structures without permission" was normalized and standardized and a total of 29,000 illegal activities corrected, with more than 8.1 million m^2 of illegal structures dismantled, more than 7,000 km of banks freed from illegal use, more than 8.1 million tons of garbage cleaned up in river courses, more than 2,700 illegal sand mining sites removed, and more than 1,000 illegal sand vessels banned, which resulted in great improvement. In addition, a clean-up action was conducted to shorelines in the mainstream of the Yangtze River, with more than 2,441 projects that violated laws and regulations removed including clean-up of 162 km of shoreline and revegetation of more than 12 million m^2 of land. Since the enactment of ten-year fishing ban on the Yangtze River, 63 illegal low embankments were found in the mainstream of the Yangtze River, Dongting and Poyang Lake cleaned up. Thanks to the removal of 59 km enclosed embankments, more than 68,000 mu of water bodies were recovered. A joint action of controlling sand mining and banning illegal sand-mining vessels was conducted by Ministry of Public Security, Ministry of Transport, Ministry of Industry and Information Technology and State Market Regulatory Administration, to Investigate and give due punishment to 1,867 violating sand-mining activities and 185 vessels (including 27 "invisible" sand-mining vessels), and turned over 104 illegal cases to the public security. Supervision and investigation were strengthened, as 3,234 rivers (5,192 sections) and 991 lakes (1,124 sections) were inspected without prior notice, covering all cities with districts under their jurisdictions in 31 provinces (autonomous regions and municipalities), rivers with a catchment area of above 1,000 km^2 and lakes of above 1 km^2 (except depopulated zone). Regarding major river basins of the Yangtze, the Yellow and the Grand Canal, a staying on-site investigation to 8 provinces were carried out.

Water resources management. Formulation of control indicators for water resources management was accelerated and a total of 63 cross-province water allocation plans were approved. The targets of securing ecological flow of 82 cross-province major rivers and lakes were organized and completed. Guidance was given to provinces to define the targets of securing ecological flow of 134 cross-province major rivers and lakes. Total amount control of groandwater intake and water level double control targets for 13 provinces (autonomous regions and municipalities) were defined. The control target of total water use in 14th Five-Year Plan period was specified and broke down to each province. A special action was taken to water abstraction, with completion of approval and register of all water intakes. Investigations were conducted to the current situation of compliance of water intakes and water-taking monitoring and metering of 5.8 million water intakes for collection of information and rectification. Inspections of water resources evaluation reports were organized for 5 national agriculture advanced and new technology industrial demonstration areas. The issue of water abstraction licenses was terminated for 13 prefectures and cities with surface water over-abstraction and 62 counties with groundwater over-abstraction. Electronic licenses are widely applied, and about 52,000 electronic licenses were issued for stock certificates. Reduction of groundwater withdrawal in water receiving areas of South-to-North Water Diversion Project were encouraged with a reduction of 3.017 billion m^3 of groundwater withdrawal. Performance evaluation for the implementation of the most stringent water resources management system in the 13th Five-Year Plan was concluded and the results were reported to the State Council for announcement after the approval. The results shall also be reported to the Organization Department of the CPC Central Committee as the key basis for comprehensive evaluation of government executives. China Water Exchange has completed a total 1,511 entitlement trading, with an amount of 308 million m^3 of water.

Integrated water resources allocation in river basins. In 2021, the Ministry of Water Resources (MWR) issued *the Management Measures for Water Resources Allocation*, which stipulates the authority, basis, organization, implementation, monitoring, supervision and management, liability and accountability of water resources allocation. In 2020 – 2021, the phase-Ⅰ of Eastern Route of South-to-

North Water Diversion transferred 674 million m³ to Shandong; the phase-I of the Middle Route of South-to-North Water Diversion transferred a total of 9.054 billion m³ of water to Beijing, Tianjin, Hebei and Henan, which great enhanced their capacity to safeguard water supply security. We channeled great energy into integrated water resources allocation in river basins, and realize integrated allocation of 31 cross-province rivers including Hanjiang River, Jialingjiang River and Wujiang River. Thanks to water allocation from the Pearl River of 17 times in dry period, the water supply in Macao and Zhuhai was secured. Integrated water allocation in major river basins was highlighted, as the mainstream of the Yellow River flows continuously for 22 years without drying up. The river and lake in North China received ecological flow on a regular basis and intensive water supply in the summer time. In 2021, a total of 8.468 billion m³ of water was transferred to 22 rivers and lakes in North China. The rivers that had dry up for many years, namely Yongding, Chaobai, Hutuo and Daqing, were recovered and flowed again. The ecosystem of Lake Wuliangsuhai is kept improving through water diversion of 598 million m³. The Lake Dongjuyanhai in the downstream of Heihe has kept running continuously for 17 years without drying up. The Yongding River entered into the sea for the first time since it dried up in 1996.

Operation and management. In 2021, the number of approved national water scenic spots reached 902, including 379 reservoirs, 202 natural rivers and lakes, 204 lake or riverine cities, 47 wetlands, 32 irrigation districts and 38 soil conservation areas.

Water pricing reform. MWR, in collaboration with the National Development and Reform Commission (NDRC), promulgated *the Action Plan for Deepening Pricing System Reform in 14th Five-Year Plan Period*, which could create a mechanism that favorites to set price of water resources structures and suits for investment and finance system, to boost water conservation and sound operation of water structures. *The revision of Measures for the Administration of Water Supply Price for Water Resources Projects* and *the Measures for Supervision and Examination of Pricing and Cost of Water Supply for Water Resources Projects* has been speed up, in order to perfect water pricing and dynamic adjustment mechanisms. By the end of 2021, the area implemented reform of agricultural water pricing accumulated to 600 million mu, with a newly increase of 160 million mu.

Water resources planning and early-stage work. In 2021, there were 38 water resources plans approved by central government authorities (including printed and issued review comments). After the approval of the State Council, the Water Security Plan of 14th Five-Year Plan Period was jointly issued by NDRC and MWR. Meanwhile, detailed and implementation plans were also worked out and issued that formed a "1+N" planning system. Together with NDRC, MWR prepared the plan and outline of national water network and reported to the State Council. The main tasks for implementing national strategy of regional development were fully committed, including regional integration development of the Yangtze River Delta, ecological protection and high-quality development of the Yellow River Basin, Chengdu-Chongqing Economic Circle, and comprehensive improvement of water environment of Taihu Basin. The process of examination and approval was accelerated for overall planning of key river basins and major tributaries, with approval of plans for Minjiang River, Hanjiang River and Lalin River. The planning of other tributaries also sped up. MWR organized to revise and prepare the flood control plans and competed the project reports. In 2021, NDRC approved 9 feasibility study reports, including 5 flood control projects and 4 water allocation projects, with a total investment of 40.703 billion Yuan. MWR approved 8 preliminary designs, including 5 flood control projects and 4 water allocation projects, with a total investment of 35.537 billion Yuan.

Soil and water conservation. In 2021, MWR formulated *Implementation Plan for Soil and Water Conservation in 14th Five-Year Plan*, issued *the Gaidance of the ministry of water Resources on Further Promoting the Construction of water and soil conservation Projects by Replacing subsidies with Awards* and *the Notice on Strengthening Water and Soil Conservation for Construction of Water Resources Projects*. MWR approved 111,900 soil and water conservation plans of construction projects, covering an area of 24,300 km^2 within the scope of responsible for control of soil and water losses. Soil conservation facilities of 47,700 construction projects completed self-check and acceptance. Supervision of man-made soil erosion during production and construction has been conducted that covers the whole country (except Hong Kong, Macao and Taiwan). According to on-site review based on interpretation of remote sensing satellite, 24,100 projects were investigated and punished because of construction without approval, damping wastes or failing to comply with the scope of responsibility for prevention and control. Demonstration areas were created for soil and water conservation at national level, with 90 demonstration areas approved. MWR selected 5 city or county as pilots for high quality development in soil and water conservation.

Rural hydropower management. By the end of 2021, the title of green small hydropower station was awarded to 870 projects in 25 provinces. Standards for safe

production and operation had been applied to hydropower stations in rural areas. The completed hydropower stations that complied with relevant standards accumulated to 3,963 in the whole country, including 99 level one, 1,498 level two and 2,176 level three hydropower stations.

Resettlement of water projects. In 2021, the resettlement population amounted to 200,900 (including 185,100 rural resettlement and 15,800 urban resettlement). Employment was arranged for 220,500 resettled people. The relocated people in rural areas, who should receive later support due to the construction of large and medium-sized reservoirs in 2020, totaled 150,200.

Supervision. In 2021, MWR organized supervision, inspection and evaluation activities based on planned schedule and agenda and matter of record, with a focus on "implementation of most stringent water resources management system", "flood and drought disaster prevention", "water project construction and operational management" and "soil and water conservation and law enforcement for water administration accountability", covering 32 matters relating to flood control, water resources management, irrigation and drainage and drinking water supply in rural areas. A total of 2,321 inspection teams with 10,221 person-times were dispatched for 21,700 projects. MWR organized "regulatory talks" or gave punishments above this level to 387 enterprises who had failed to comply with the relevant stipulations. MWR organized a three year special rectification action on safe production and a special action for hidden danger rectification and investigation, with a coverage of 99.6% and 2,550 rules and regulations promulgated. Production safety patrols and major quality and safety inspection were strengthened for key areas and links related to project construction and operation. MWR issued 46 rectification comments by means of "one province one order", and held "regulatory talk" with 5 persons-in-charge who held responsibilities. Great attention has been paid on standardization of safe production. In 2021, a total of 131 enterprises passed the appraisal of Grade I compliance. MWR initiated and advanced the action of "safety supervision + informatization". In the fourth quarter of 2021, the national average safety risk was 14.26—45.8% lower than that in 2020—which means the overall risk was controllable. MWR dispatched 75 person times in production safety patrols and major quality and safety inspection tours and identified 162 problems with quality safety, and issued 13 rectification notices by means of "one province one order". More attention has been paid on major works of flood control in river basins and national water network, with 7 batches of 73 groups and 675 person times, involving 127 major water resources projects and risk reduction and strengthening of reservoirs. MWR issued 58 rectification notices by means of "one province one order", and held "regulatory talk" with 86 responsible organizations. MWR urged the provincial water departments to conduct self-inspection, with 150 batches, 460 inspection groups, nearly 3,900 person times and 1,070 projects.

Legislation and administrative law enforcement. In 2021, a water-related law and a administrative regulations were enacted. MWR completed drafting of 3 laws and regulations. In 2021, the number of water-related cases investigated totaled 19,800 and 18,600 cases or 93.9% were resolved. MWR handled and concluded 12 administrative reconsideration cases and 20 administrative proceedings.

Administrative permits. In 2021, MWR handled 2,561 applications for water-related administrative approvals or permits with 1,727 completed, including examination and verification of 39 water project construction plans, approval of 3 water projects on the boundary of different administrative regions, approval of 8 preliminary design reports of water construction projects, 154 water abstraction licenses, 31 evaluation reports of flood impact by non-flood control project, 396 plans of construction projects within the jurisdiction of river courses, 64 approvals of soil and water conservation plan of production and construction projects, 3 approvals for establishment and adjustment of national basic hydrological stations, 113 approvals of hydrological monitoring projects for evaluating impact of construction at upper and lower of the national basic hydrological stations, 1,368 qualification approvals (including new application, adding of new items or promotion) for construction supervisors of water resources projects; and 415 Grade-A qualification identifications (including new application and extension) for quality supervisors of water-related projects.

Water science and technology. In 2021, the central government approved and allocated a total of 404 million Yuan to water-related science and technology projects, including 23.855 million Yuan for 9 special-subject water-related projects listed in the National Key Research and Development Plan namely "Comprehensive improvement of water Resources and water Environment in major river basins of the Yangtze River and the Yellow River" and "Prevention and control of major natural disasters and public safety"; 14.8386 million Yuan for 48 project with joint funds for scientific studies on the Yangtze River and the Yellow River; 15 million Yuan for 58 water science and technologies demonstration projects and 55 of which were filling for conclusion. One MWR project won the second prize of State Science and Technology Awards. By the end of 2021, there were 17 national level or ministerial level labs (including 5 labs in preparation stage), 15 national and ministerial engineering technology research centers. A total of 94.01 million Yuan was allocated from the central government as operation expenses for basic scientific studies of public research institutions. There were 34 water-related technical standards made public and 145 standards under drafting. The number of effective water-related technical standards amounted to 802.

International cooperation. In 2021, an agreement on international cooperation was singed. MWR organized 35 multi-bilateral round-table meetings or technical exchange seminars, and participated 50 online conferences organized by water-related international organizations. 8 projects with the loans or grants from the World Bank, the Asian Development Bank and Global Environmental Funds made smooth progress. Steady progress has been made by Sino-Sweden, Sino-Demark, Sino-France and Sino-Finland cooperation and international scientific cooperation projects. The outcome of the third Lancang-Mekong Cooperation (LMC) Leaders' Meeting was full implemented. The 2nd Lancang-Mekong Water Resources Cooperation Forum was held successfully. A total of 9 Asia Cooperation Fund Projects were organized for implementation, with a funding of 19.63 million Yuan. In order to further up cooperation for flood control, we strengthened hydrological information exchange and sharing of scheduling information of reservoirs in the tributaries with the Russia for emergency purpose. The neighboring countries received flood warning and reporting from 72 hydrological stations in China.

VII. Current Status of the Water Sector

Water-related institutions. By the end of 2021, there were 20,413 legal entities, with 865,000 employees and separate accounts, engaged in water-related activities. Among them, the number of governmental organizations were 2,704 with 125,000 employees, up by 0.8% over the previous year; public organizations were 13,729 with 459,000 employees, down by 5%; enterprises were 3,332 with 273,000 employees, down by 5.2%; societies and other institutions were 648 with 6,000 employees, up by 50%.

Employees and salaries. Employees of the water sector totaled 779,000, down 3.72% from the previous year. Of which, in-service staff members amounted to 748,000, down 3.81%, including 60,000 working in agencies directly under the Ministry of Water Resources, down 9.76% over the previous year; and 688,000 working in local agencies, down 3.25%. The total salary of in-service staff members nationwide was 81.87 billion Yuan, and the annual average salary per person was 110,000 Yuan.

Employees and Salaries

	2011	2012	2013	2014	2015	2016	2017	2018	2019	2020	2021
Number of in service staff/10^4 persons	102.5	103.4	100.5	97.1	94.7	92.5	90.4	87.9	82.7	77.8	74.8
Of them: staff of MWR and agencies under MWR/10^4 persons	7.5	7.4	7.0	6.7	6.6	6.4	6.4	6.6	6.6	6.7	6.0
Local agencies/10^4 persons	95.0	96.0	93.5	90.4	88.1	86.1	84.0	81.3	76.0	71.1	68.8
Salary of in-service staff/10^8 Yuan	351.4	389.1	415.3	451.4	529.4	640.5	739.1	802.7	787.6	790.9	818.7
Average salary/(Yuan/person)	34,283	37,692	41,453	46,569	55,870	69,377	83,534	91,307	95,385	102,000	110,000

Main Indicators of National Water Resources Development (2016–2021)

Indicators	unit	2016	2017	2018	2019	2020	2021
1. Irrigated area	10^3 ha	73,177	73,946	74,542	75,034	75,687	78,315
2. Farmland irrigated area	10^3 ha	67,141	67,816	68,272	68,679	69,161	69,609
Newly-increased in 2020	10^3 ha	1,561	1,070	828	780	870	1,114
3. Water-saving irrigated area	10^3 ha	32,847	34,319	36,135	37,059	37,796	
Highly-efficient water-saving irrigated area	10^3 ha	19,405	20,551	21,903	22,640	23,190	
4. Irrigation districts over 10,000 mu	unit	7,806	7,839	7,881	7,884	7,713	7,326
Irrigation districts over 300,000 mu	unit	458	458	461	460	454	450
Farmland irrigated areas in irrigation districts over 10,000 mu	10^3 ha	33,045	33,262	33,324	33,501	33,638	35,499
Farmland irrigated areas in irrigation districts over 300,000 mu	10^3 ha	17,765	17,840	17,799	17,994	17,822	17,868
5. Rural population accessible to safe drinking water	%	79	80	81	82	83	84
6. Flooded or waterlogging area under control	10^3 ha	23,067	23,824	24,262	24,530	24,586	24,480

Continued

Indicators	unit	2016	2017	2018	2019	2020	2021
7. Controlled or improved eroded area	10^4 km^2	120.4	125.8	131.5	137.3	143.1	149.6
Newly-increased	10^4 km^2	5.6	5.9	6.4	6.7	6.4	6.8
8. Reservoirs	unit	98,460	98,795	98,822	98,112	97,072	97,036
Large-sized	unit	720	732	736	744	756	805
Medium-sized	unit	3,890	3,934	3,954	3,978	4,043	4,174
Total storage capacity	10^8 m^3	8,967	9,035	8,953	8,983	9,086	9,853
Large-sized	10^8 m^3	7,166	7,210	7,117	7,150	7,231	7,944
Medium-sized	10^8 m^3	1,096	1,117	1,126	1,127	1,146	1,197
9. Total water supply capacity of water projects in a year	10^8 m^3	6,040	6,043	6,016	6,021	5,813	5,920
10. Length of dikes and embankments	10^4 km	29.9	30.6	31.2	32.0	32.8	33.1
Cultivated land under protection	10^3 km	41,087	40,946	41,351	41,903	42,168	42,192
Population under protection	10^4 people	59,468	60,557	62,785	67,204	64,591	65,193
11. Total water gates	unit	105,283	103,878	104,403	103,575	103,474	100,321
Large-sized	unit	892	892	897	892	914	923
12. Total installed capacity by the end of the year	10^4 kW	33,153	34,168	35,226	35,564	36,972	39,184
Yearly power generation	10^8 kW·h	11,815	11,967	12,329	12,991	13,540	13,419
13. Installed capacity of rural hydropower by the end of the year	10^4 kW	7,791	7,927	8,044	8,144	8,134	8,290
Yearly power generation	10^8 kW·h	2,682	2,477	2,346	2,533	2,424	2,241
14. Completed investment of water projects	10^8 Yuan	6,099.6	7,132.4	6,602.6	6,711.7	8,181.7	7,576.0
Divided by different sources							
(1) Central government investment	10^8 Yuan	1,679.2	1,757.1	1,752.7	1,751.1	1,786.9	1,708.6
(2) Local government investment	10^8 Yuan	2,898.2	3,578.2	3,259.6	3,487.9	4,847.8	4,236.8
(3) Domestic loan	10^8 Yuan	879.6	925.8	752.5	636.3	614.0	698.9
(4) Foreign funds	10^8 Yuan	7.0	8.0	4.9	5.7	10.7	8.1
(5) Enterprises and private investment	10^8 Yuan	424.7	600.8	565.1	588.0	690.4	718.2

Continued

Indicators	unit	2016	2017	2018	2019	2020	2021
(6) Bonds	10^8 Yuan	3.8	26.5	41.6	10.0	87.2	104.3
(7) Other sources	10^8 Yuan	207.1	235.9	226.3	232.8	144.9	101.1
Divided by different purposes:							
(1) Flood control	10^8 Yuan	2,077.0	2,438.8	2,175.4	2,289.8	2,801.8	2,497.0
(2) Water resources	10^8 Yuan	2,585.2	2,704.9	2,550.0	2,448.3	3,076.7	2,866.4
(3) Soil and water conservation and ecological recovery	10^8 Yuan	403.7	682.6	741.4	913.4	1,220.9	1,123.6
(4) Hydropower	10^8 Yuan	166.6	145.8	121.0	106.7	92.4	78.8
(5) Capacity building	10^8 Yuan	56.9	31.5	47.0	63.4	85.2	79.9
(6) Early-stage work	10^8 Yuan	174.0	181.2	132.0	132.7	157.3	136.5
(7) Others	10^8 Yuan	636.2	947.5	835.8	757.4	747.3	793.8

Notes:

1. The data in this bulletin do not include those of Hong Kong, Macao and Taiwan.

2. Key indicators for water development and statistical data in 2012 and in 2013 is also integrated with the data of first national census for water.

3. Statistics of rural hydropower refer to the hydropower stations with an installed capacity of 50,000 kW or lower than 50,000 kW.

《2021年全国水利发展统计公报》编辑委员会

主　　　　任：刘伟平
副　主　　任：吴文庆　张祥伟
委　　　　员：（按姓氏笔画排序）
　　　　　　匡尚富　邢援越　巩劲标　朱　涛　朱闽丰　任骁军
　　　　　　刘宝军　孙　卫　李　烽　李兴学　李原园　杨卫忠
　　　　　　吴　强　张新玉　陈茂山　郑红星　荆茂涛　姜成山
　　　　　　夏海霞　钱　峰　倪　莉　倪文进　徐永田　郭孟卓
　　　　　　曹纪文　曹淑敏　谢义彬

《2021年全国水利发展统计公报》主要编辑人员

主　　　　编：张祥伟
副　主　　编：谢义彬　吴　强
执　行　编　辑：汪习文　张光锦　李　淼　张　岚
主要参编人员：（按姓氏笔画排序）
　　　　　　万玉倩　王　超　王小娜　王鹏悦　毕守海　曲　鹏
　　　　　　吕　烨　刘　品　米双姣　孙宇飞　杜崇玲　李天良
　　　　　　李云成　李笑一　杨乐乐　吴泽斌　吴海兵　吴梦莹
　　　　　　沈东亮　张晓兰　张慧萌　周哲宇　房　蒙　郭　悦
　　　　　　黄藏青　戚　波　盛　晴　谢雨轩　潘利业
英　文　翻　译：谷丽雅　侯小虎　张林若

◎ 主编单位
水利部规划计划司

◎ 协编单位
水利部发展研究中心

◎ 参编单位
水利部办公厅
水利部政策法规司
水利部财务司
水利部人事司
水利部水资源管理司
全国节约用水办公室
水利部水利工程建设司
水利部运行管理司
水利部河湖管理司
水利部水土保持司
水利部农村水利水电司
水利部水库移民司
水利部监督司
水利部水旱灾害防御司
水利部水文司
水利部三峡工程管理司
水利部南水北调工程管理司
水利部调水管理司
水利部国际合作与科技司
水利部综合事业局
水利部信息中心
水利部水利水电规划设计总院
中国水利水电科学研究院